The United States ... locked in a "cold tech ... ner will end up dominating the twenty-first century.

Beijing was not considered a tech contender a decade ago. Now, some call it a leader. America is already behind in critical areas.

It is no surprise how Chinese leaders made their regime a tech powerhouse. They first developed and then implemented multi-year plans and projects, adopting a determined, methodical, and disciplined approach. As a result, China's political leaders and their army of technocrats could soon possess the technologies of tomorrow.

America can still catch up. Unfortunately, Americans, focused on other matters, are not meeting the challenges China presents. A whole-of-society mobilization will be necessary for the U.S. to regain what it once had: control of cutting-edge technologies. This is how America got to the moon, and this is the key to winning this century.

Americans may not like the fact that they're

once again in a Cold War–type struggle, but they will either adjust to that reality or get left behind.

5G and the Internet of Things

Nowhere is America so far behind China as in the race to build the world's next – the fifth – generation of wireless telecommunications networks.

"Not since the invention of gunpowder has China led the world in the introduction of a disruptive new technology, and the United States still can't believe that it has been leapfrogged," wrote David Goldman, the American writer and thinker, in *Tablet* in March 2019, referring to 5G. "For years the conventional wisdom held that the Chinese only could copy but not innovate. That wisdom has now been proven wrong."

The Chinese have raced ahead in 5G in large part because they made it "a central plank" of their industrial planning process, including it in both the 13th Five-Year Plan,

> *A whole-of-society mobilization will be necessary for the U.S. to regain control of cutting-edge technologies.*

which covers the half-decade ending in 2020, and the Made in China 2025 initiative. Chinese technocrats announced the addition of 5G to CM2025, as the now-notorious plan is known in China, in January 2018. Chinese leader Xi Jinping also made 5G a part of his Belt and Road Initiative when in May 2017 he announced the "Digital Silk Road." Wireless will feature prominently in the 14th Five-Year Plan, on the drafting board now.

There is a prize for the country controlling tomorrow's wireless communication networks. According to forecaster Stratfor, 5G is nothing less than "the technology that will drive the world's economy in the decade to come."

That bold assessment is obviously correct:

5G, due to speeds 2,000 times faster than existing 4G networks, will permit near-universal connectivity. Homes, vehicles, machines, robots, and just about everything else will be linked and communicating with each other. That's what is now called the Internet of Things.

Imagine a world where Beijing is connected to most devices around the planet. That gives China, already "the new OPEC of data," access to even more of it.

And by hook or by crook the Chinese will take the world's information. Huawei Technologies, as Senator Marsha Blackburn, the Tennessee Republican, told Fox News Channel in July 2019, is Beijing's "mechanism for spying."

She's right. The company, whose name translates as "For China," is in no position to refuse Beijing's demands to gather intelligence. For one thing, Beijing owns almost all of Huawei. The Shenzhen-based enterprise maintains it is "employee-owned," but that is an exaggeration. Founder Ren Zhengfei holds a 1 percent stake, and the remainder is effectively controlled by the state through a "trade union committee."

Moreover, in the Communist Party's top-down system virtually no one can resist a command from the ruling organization. The Party's power is even codified. Articles 7 and 14 of China's National Intelligence Law, enacted in 2017, require Chinese nationals and entities to spy if relevant authorities make a demand. Ren has repeatedly maintained that Huawei would never snoop, but this defiant claim, in view of everything, is not credible.

The spying concern is not theoretical. In fact, from 2012 to 2017, China surreptitiously downloaded data nightly – through Huawei servers – from the Chinese-built and Beijing-donated headquarters of the African Union in Addis Ababa.

Not surprisingly, Huawei has been positioning itself to seize tomorrow's data. First, academic Christopher Balding's study of resumes of Huawei employees reveals that some of them claim concurrent links with units of the People's Liberation Army, the Chinese military, in roles that apparently involve intelligence collection. As he writes in his July 2019 study, "there is an undeniable relationship between Huawei and the Chinese state, military, and intelligence gathering services." Founder Ren Zhengfei was an officer in the People's Liberation Army before being demobilized in 1983. He is a member of the Communist Party.

Second, recent analyses show Huawei software to contain an abnormally high number of security flaws. Finite State, a cybersecurity firm, revealed that 55 percent of nearly 10,000 scanned Huawei firmware images contained at least one backdoor vulnerability. The Chinese company's products, according to the survey, contained the most such flaws among its competitors.

The concern is not only exfiltration of information. Beijing will undoubtedly use Huawei to control the networks operating the devices of tomorrow, remotely manipulating everything hooked up to the Internet of

Things, in other words, just about everything.

With devices around the planet networked to China, Beijing could have the ability to drive cars off cliffs, unlock front doors, and turn off or speed up pacemakers. In the first moments of a war, Beijing could literally see into most corners of the world and paralyze critical infrastructure.

Who wants to give a militant one-party state the power to surveil and control internet-connected devices plus win wars? The answer, unfortunately, is many nations, even some friends of the United States. The Philippines, a U.S. treaty partner, has decided to buy Huawei 5G gear, and Italy, a NATO ally, is almost certainly going to make the same decision soon. The U.K., considered America's closest ally, announced it too will be purchasing Huawei 5G equipment.

Moreover, America will soon be surrounded. Huawei is building Mexico's 5G network, and Canada, although one of America's "Five Eyes" intelligence-sharing partners, will probably go with Huawei. Huawei claims that, by the middle of 2019, it had wrapped up 50 commercial 5G contracts.

President Trump's campaign to convince countries to not buy Huawei 5G gear, according to national security analyst Eli Lake, has already "collapsed." "Huawei," Goldman told the London Center for Policy Research in

---

Nowhere is America so far behind China as in the race to build the world's next – the fifth – generation of wireless telecommunications networks.

August 2019, "is rolling out 5G across the whole Eurasian continent – I know of not one exception, not even the U.K., and not even India."

The Global South, he believes, "will be hard-wired" into Huawei and therefore into the Chinese economy. The firm's practice of selling to developing countries resembles Mao Zedong's take-the-countryside-and-then-surround-the-cities tactic, and it would leave America, Goldman believes, like Britain after the dissolution of its empire.

In sum, as Dimitris Mavrakis of ABI Research told *The Wall Street Journal*, "5G will be made in China."

There's a reason for the Chinese company's success. Huawei offers equipment at costs far below that of competitors. Some put the Huawei discount at 20 percent, leaving competitors Ericsson of Sweden and Nokia of Finland at an almost insurmountable disadvantage. In some cases, Huawei gear is as much as 30 percent cheaper, and sometimes it pitches 40 percent below the next-cheapest bid.

The reason the Chinese company, the world's largest supplier of telecom-networking gear, can offer such low prices can be boiled down to two words: subsidies and stealing.

As essentially a state-owned enterprise – and certainly as a "national champion" – Huawei has been the beneficiary of generous subsidies from the Chinese central government, perhaps as much as $75 billion according to a *Wall Street Journal* investigation. The already large subsidies are thought to have increased substantially in 2011 and again in 2018. The increase in 2018 was triggered by the realization how dependent China was on American technology following the Trump administration's decision – quickly reversed – to cut-off a sister company, Shenzhen-based ZTE Corp., from U.S. technology.

And then there is outright theft from American and other companies. Since just about the moment it was formed in 1987, Huawei has been implicated in stealing technology, from Cisco Systems and others. The theft was so pervasive that Huawei drove out foreign competition. It is often blamed for killing off, most notably, Canada's Nortel Networks.

Huawei, according to recent allegations, has never stopped stealing for product-development purposes. The U.S. Justice Department in January 2019 unsealed an indictment against the company charging it with taking unauthorized photographs and an arm of "Tappy," a cellphone-testing robot of T-Mobile. The FBI, according to Bloomberg reporting, is investigating Huawei for pilfering the technology for advanced smartphone glass from Akhan Semiconductor, a small Illinois-based firm. The *Wall Street Journal* reported in August 2019 that U.S. prosecutors are continuing their investigation of the Chinese telecom company for intellectual property theft.

At the moment, there is no American telecom-equipment giant, no "American Huawei," as one writer put it not long ago. The fall of U.S. competitors is all the more striking because the technology was first developed in America – Motorola, before broken into parts, engineered the world's first cell call – and not long ago American companies like AT&T sat atop the rankings for telecom-equipment makers.

There were many reasons for the decline of equipment manufacturing in America.

Huawei's rise did not help of course, but U.S. companies also had themselves to blame. They chased high stock market valuations by deciding to get out of low-margin manufacturing. Their strategy was to concentrate on capturing the richest profits in wireless communications, licensing the technology itself.

Those big margins belong to the company that once dominated 3G and now essentially owns 4G, Qualcomm Inc. The San Diego-based giant designs and produces the chips that make the current wireless networks work. So if you buy a Huawei smartphone – it is the world's No. 2 maker of those devices – chances are it has a Qualcomm chip inside. That's also true for just about every other cellphone or mobile device.

But what about 5G? Qualcomm's ability to extend its reach to the future is under attack – from Washington regulators. In May 2019, Judge Lucy Koh of the Federal District Court in San Jose sided with the Federal Trade Commission and against Qualcomm in an antitrust case the Commission brought three days before the inauguration of Donald Trump.

The FTC, in its action against Qualcomm, alleged that the company used dominant market power in modem chips to charge royalties that were too high, likening "elevated royalties" to a "tax." The FTC's novel theory would, if ultimately accepted, make the Federal government a de facto price regulator setting mobile technology royalty rates. As such, the Commission essentially introduced price control and attacked Qualcomm's famed business model of using licensing revenue to fund innovation.

The apparent desire to protect Qualcomm's

5G is nothing less than "the technology that will drive the world's economy in the decade to come."

innovation model led Trump to issue an unprecedented presidential order in March 2018 blocking Broadcom Ltd. from taking steps to acquire the San Diego–based firm. The concern was that Broadcom, if it were successful in its $117 billion bid, would break Qualcomm into parts to realize locked-in shareholder value and thus cripple the firm's ability to compete with China in 5G.

Qualcomm, as a practical matter, is the only American business that can lead the world in 5G. "I can make a case that if Qualcomm's 5G motivations were to wane, the U.S.'s ability to lead the world in 5G wanes," writes Patrick Moorhead, a *Forbes* contributor. "Sure, there are other important, American companies engaged in 5G, but I believe none who can move the world as quickly and comprehensively and solve the biggest wireless problems as Qualcomm."

That assessment, which appears to be exactly correct, suggests that the Federal District judge's decision in favor of the FTC was a gift to China. The Chinese themselves must think so. On the first day of the Commission's case in Federal court, its star witnesses were from China, Lenovo Group, and the infamous Huawei Technologies.

So if the FTC ultimately gets its way – the case is on appeal – China's Huawei, not America's Qualcomm, will dominate 5G.

Americans need to know the consequence of that. "China's game," David Goldman states, "is to control the broadband, and then the e-commerce, and then the e-finance, and then all the tech startups servicing the 'ecosystem,' and then the logistics." As he told me, "The world will become a Chinese company store."

ARTIFICIAL INTELLIGENCE

China, writes Amy Webb in *Inc.*, has been "building a global artificial intelligence empire, and seeding the tech ecosystem of the future." Webb, the founder of the Future Today Institute, believes it has been particularly successful. She tells us that "China is poised to become its undisputed global leader, and that will affect every business."

Webb's assessment is now commonly accepted. "China's advantages in size, data collection, and national determination have allowed it, over the past decade, to close the gap with American leaders of this industry," wrote Graham Allison of Harvard's Kennedy School in December 2019. "It is currently on a trajectory to overtake the United States in the decade ahead."

Artificial intelligence permits machines to mimic human functions. Powered by AI, they can drive vehicles, recognize spoken words, and play games of skill like chess and Go.

Especially Go, the Chinese game of strategy. If China had a "Sputnik moment," it occurred in March 2016 when AlphaGo, developed by Alphabet Inc.'s DeepMind Technologies, took four out of five games from an 18-time champion in a challenge match in Seoul.

By the following year, Chinese officials were pouring even more money into AI research. Beijing in 2017 supplemented the AI component of Made in China 2025 with its "Next Generation Artificial Intelligence Development Plan," a three-part effort to lead global AI by 2030.

Furthermore, Beijing made sure its determination to dominate artificial intelligence was matched across Chinese society. Business chieftains and policy analysts in China are much more focused on AI than those in the West, surveys show.

As Webb indicates, Beijing's nationwide effort paid off. China, for instance, now publishes more machine learning papers than the United States. Moreover, China's share of the top 10 percent most-cited papers is rising, hitting 26.5 percent in 2018. That compares to America's 29.0 percent, a percentage on the decline. China's average citations are also up, behind the U.S. but nonetheless above the world average.

Whichever nation wins at AI will secure important leads in both the global economy and conventional military warfare. To borrow a phrase from Hollywood, we are witnessing the "Rise of the Machines."

What if the "machines" are Chinese? The world is getting an ominous look at what will

---

With devices around the planet networked to China, Beijing could have the ability to drive cars off cliffs, unlock front doors, and turn off or speed up pacemakers.

---

occur in what Beijing euphemistically calls the Xinjiang Uygur Autonomous Region. Facial recognition systems, powered by AI, are helping China's leaders continually track the region's inhabitants.

In Xinjiang, Beijing is relentlessly eliminating cultural and religious identity and implementing race-based policies reminiscent of those of the early Third Reich. For example, more than a million Xinjiang inhabitants are being held in concentration-camp-like facilities for no reason other than their minority – Uighur or Kazakh – ethnicity or their adherence to Islam. Children are kept in prison-like "orphanages."

Unfortunately, American companies are helping China's leaders in what many label – correctly – crimes against humanity. For instance, AI researchers from Microsoft, Rensselaer Polytechnic Institute, and Michigan State University gave keynote speeches at the Chinese Conference on Biometric Recognition in Xinjiang in August 2018 on facial recognition, used by Beijing for hideous social-control purposes.

China is on the AI map in part because Beijing has been given a boost by U.S. companies sharing their technology. Leaders in the field are both Alphabet and its Google unit. Alphabet is a major player in part due to its acquisition of London-based DeepMind in 2014. Google conducts extensive AI research separate from its parent company.

Some of Google's research is in China.

The company has three AI operations there: the Google AI China Center in Beijing, established in 2017, and partnerships with China's two premier educational institutions, Peking University and Tsinghua University.

At the same time, Google has shunned cooperation with the Pentagon. Peter Thiel, the Silicon Valley investor, has severely criticized the search giant. "I think it is unprecedented in the last 100 years, or ever, that a major U.S. company refused to work with the U.S. military and has worked with our geopolitical rival," he said on Fox News Channel's "Sunday Morning Futures" in August 2019.

Google has in various statements denied charges like the ones Thiel has been making, but the company's contentions, although technically true, are highly deceptive. First, the company has said it works with the Pentagon, but it is nonetheless not renewing its Project Maven contract, an AI project analyzing drone footage.

Second, Google denies working with the Chinese military, but in the China of Xi Jinping, "civil-military fusion" means nominally civilian research is quickly pipelined into the Chinese military. In any event, the Communist Party, to which the People's Liberation Army reports, has, in reality, near-absolute power over something as important as scientific and technical research. Companies like Google in fact know – or do not want to know – about the military's access to their AI research in China.

Not everyone is concerned about China's militarization of research. "You're not going to be able to stop or slow down Chinese AI progress by stopping these labs," Jeffrey Ding of Oxford's Center for the Governance of AI told Vox, referring to foreign AI research facilities. "Either we try to get the best and brightest, or they have other options."

Ding highlights an important aspect of the AI race. The competition, as a practical matter, is one for people. As futurist George Gilder has noted, "The most precious resource in the world economy is human genius." America should know Gilder is right. Axios reports that most of the best AI researchers in the United States came from elsewhere.

"What has given the US its AI advantage has been, in significant part, the fact that the U.S. attracts AI talent from all over the world," Vox writes. "While America is a much smaller country than China, it's drawing on what is effectively a much larger talent pool, including attracting many top Chinese researchers." According to a 2018 report, China ranks second in the number of AI scientists and engineers, coming in at 18,200. The U.S. was first with about 29,000.

The ability to attract and retain people is China's big vulnerability. Beijing is now keeping more homegrown talent, in part because it has better retainment programs and because friction with the U.S. makes that country less attractive for Chinese researchers. China is also offering rich salaries. According to an Ohio State University study published in December 2019, the country is getting more of its scientists to return home. Yet despite everything, Beijing's steps are hardly sufficient. As the title of an August 2019 article on the website of *MIT Technology Review* tells us, "China's Path to AI Domination Has a Problem: Brain Drain."

Although much AI research today is open-source – meaning it does not matter where researchers are based – in coming years AI work will not, in all probability, be published in public forums. That should put a premium on attracting the best talent to one's own country. Moreover, in a competitive era, the closing off of AI work product will surely be mimicked in other areas.

The increasingly restrictive atmosphere throughout China, the result of Xi Jinping's greater emphasis on ideological rectitude, does not necessarily mean the country cannot be a scientific or technical leader. China has made great strides in various areas, but it has had to import technicians, researchers, and scientists to do so. In an increasingly xenophobic era – Xi Jinping promotes "Han nationalism," a race-based ideology – it is unlikely to keep that foreign talent over the long term.

Moreover, Xi's demand for absolute political obedience suffocates society. Suffocation can only inhibit critical thinking, an essential element of research, even in areas far removed from Xi's Maoist-like ideology.

There is, however, an even more important "general atmospherics" problem. China will have to build the trust necessary to gain users from around the world for its products. During his tenure, however, Xi has shown a disturbing tendency to employ all the means at his disposal to achieve internal political and diplomatic objectives. Beijing has been busy trying to establish international standards for the development and use of artificial

In the first moments of a war, Beijing could literally see into most corners of the world and paralyze critical infrastructure.

intelligence and is in the lead in setting them, but it's hard to see how any communist system, and especially Xi's, can adhere to them.

If China does not adhere to standards, however and wherever they are set, it will be difficult for its researchers to build collaborations with those elsewhere and, in the absence of outright theft, to gain access to foreign data. As Joy Dantong Ma of the Chicago-based Paulson Institute told *Nature*, "It's in their interest to play fair."

This is true, but it is not in the nature of their political system – or their leader of the moment – to do so.

QUANTUM

There is a "Quantum Revolution" coming. As the *Washington Post* reports, Chinese scientists "are at the forefront." China's research advances in quantum have been breathtaking.

The weird behavior of atoms and subatomic particles can give rise to previously unimaginable computing power and hack-proof communications. Beijing leaders have recognized the civilian and military potential, pouring immense resources into the areas.

The determined effort is paying off. China is clearly in the lead when it comes to taking advantage of the communications applications of what Albert Einstein famously described as "spooky action at a distance," the phenomenon of separated particles moving in coordinated fashion. In 2016, China launched a satellite and, harnessing the power of this "entanglement," soon conducted a video conference call between Beijing and Vienna.

Now, China plans an array of satellites and fiber-optic cable to build a secure quantum communications network. Already, 1,300 miles of the network, including links connecting Beijing and Shanghai, are operating. A Chinese quantum satellite, in the first event of its kind, linked up with a mobile quantum ground station for an encrypted eight-minute transmission in late December 2019.

In, say, three years, American spies will no longer be able to eavesdrop on Chinese communications. Tapping into a line, however accomplished, will break the "spooky" movement of particles and therefore leave the intruder with nothing to take or use.

In the communications sector, China's lead is long. In 2018, the country registered 517 quantum communications and cryptography patents. The U.S. came in a distant second with 117. Europe had 31. No wonder China has been called a "Subatomic Superpower."

In the more important area of quantum computing, however, China is playing catch-up to America. Google has what is considered the most advanced quantum computer in the world, a 72-quantum-bit machine. IBM's computer is 50 qubits. China is "working on 24 qubits," according to Zhu Xiaobo of the University of Science and Technology of China.

China, far behind, has plenty of time to make up ground, however. It will take at least a decade – and perhaps two of them – before the first full-scale quantum computer hits the market. Chinese officials, from Xi Jinping on down, are placing big bets on quantum and devoting the resources to overtake America.

The center of China's innovation will be Hefei, the capital of Anhui province. The National Laboratory for Quantum Information Sciences, a multi-billion-dollar facility, is spread over 86 acres, and when fully operational will bring researchers from across China together. "This may sound a bit old-fashioned, even Soviet-style, but it can give China a chance to win the race," said Guo Guoping of the Chinese Academy of Sciences, which will run the location, the world's largest quantum research lab.

Some, questioning the concept of a national lab, think it's not a good idea to concentrate the nation's quantum work in one place. Others believe the "enormous bet" on quantum research is not smart in the first place. "Critics have warned that spending more taxpayers' money on quantum research will only result in less funding for other disciplines," reports Hong Kong's *South China Morning Post.*

"I have known some researchers whose funding applications were turned down because their study was unrelated to quantum," said a professor at the Beijing University of Posts and Telecommunications, speaking without attribution to the paper. "I have also met officials who knew nothing about quantum physics but took the bait on quantum computers and networks as offering a solution to all problems."

The lab nonetheless has the backing of China's "father of quantum," Pan Jianwei, who is based in the University of Science and Technology of China. USTC, as the institution is known, will have a large role in the lab, which will also have links to the People's Liberation Army. According to the D.C.–based Center for a New American Security, USTC maintains quantum research efforts with state-owned defense enterprises. Pan, who personally maintains military links, told the *Anhui Business Daily* that the new facility will be immediately useful to the Chinese military.

The national lab also appears to be of immediate use to foreign business. Brandon Weichert, who publishes *The Weichert Report*, told radio host John Batchelor in October 2018 that Google has expressed interest in investing in and contributing personnel to the new lab.

Google's AI efforts in China, Weichert says, will require the company to expand its cloud computing capabilities in the country – it is exploring cooperation with China's Tencent in this area – and that in turn means it will also need quantum computing.

Yet cooperation with China raises national security issues for the United States. As Weichert told me, "China has been anticipat-

---

Who wants to give a militant one-party state the power to surveil and control internet-connected devices plus win wars?

---

ing this for a few years and they are making moves to coopt Google research on quantum as best as they can, just as they did with Western manufacturing firms 20 years ago."

"The Chinese," he said, "are replicating the very same behavior they used to gut the manufacturing sector from the United States, and they are applying those techniques and capabilities in gutting America's high-tech sector."

In response, the United States is beginning to restrict American cooperation with China. The Department of Energy, for instance, in June 2019 banned employees and contractors from participating in some foreign recruiting efforts, such as Beijing's "Thousand Talents" program. As the American government begins to pour money into quantum – the recently enacted National Quantum Initiative Act authorizes an additional $1.2 billion for quantum research – researchers will tend to stay home to keep Federal grants.

Where talent goes determines the race for quantum, and the stakes are high. Quantum, as some say, "has the potential to upset the geopolitical balance of power." Because it does, America must be at the forefront. "Whatever the quantum age looks like, one thing seems certain: it will dawn," writes *GQ*'s Charlie Burton. "The nations that define this century will be there when it does."

WHAT TO DO

The shape of America's response to China's technological advances will depend on the nature of the overall competition with Beijing. Is the U.S. merely competing against another large state or is it facing an implacable foe seeking its extinction?

Analysts are fond of saying that China and the United States are involved in just one of history's many boys-will-be-boys contests for dominance.

This assessment implies America is jealously attempting to preserve its position atop the international system. That is Beijing's narrative, as Chinese leaders denigrate America by saying it is on the way down and trying to prevent their country's legitimate rise.

In reality, Americans are preserving more than just their role in the international system. They are trying to preserve that system itself.

Xi Jinping unfortunately does not accept that system of sovereign states competing in a network of treaties, conventions, rules, and norms. He is, on the contrary, attempting to impose on the world China's imperial-era concepts, including the notion that underpinned the tributary system. Chinese emperors, believing they were predestined and compelled to order and rule the entire world, claimed they possessed the Mandate of Heaven over *tianxia*, or "All Under Heaven."

Xi has employed *tianxia* language for more than a decade, but recently his references have become unmistakable as have those of his subordinates, and by now it is clear he thinks his China is the world's only sovereign state. The trend of Xi's recent comments warns us he does not want to live within the current Westphalian international system or even to adjust it. Xi, from every indication, is hoping to overthrow it altogether.

If this breathtaking, revolutionary challenge were not bad enough, Xi's People's Republic is a militant state that considers the United States its enemy. That hostile view looks to be official. In May 2019, for instance, *People's Daily*, the Communist Party's "mouthpiece" and therefore most authoritative publication in China, and Xinhua News Agency, the central government's official media outlet, carried a piece declaring a "people's war" on America.

With the challenge to America so fundamental, it becomes imperative for Washington to win the tech challenge so that it can fend off a danger not only to itself but also to the international system. Therefore, Americans need to ask how to prevent the world's most dangerous regime from dominating the world's most powerful technologies. And they need to be prepared to take drastic, emergency-like measures to prevent Chinese success.

Beijing's efforts to develop key technologies have been almost entirely massive, state-directed, and government-funded. As noted, China's top-down model has worked, especially in the 5G area. The Chinese economic model, David Goldman has stated, is "a souped-up, bigger, and more ruthless version" of what Japan's Meiji emperor did after the restoration in 1868. Since then, other Asian nations – most notably South Korea and Taiwan – have copied it, albeit with local adaptation.

China has made the Japanese approach work, for instance driving American companies out of capital-intensive manufacturing. Beijing's success raises a question: Should America mimic the Japanese-Chinese-Taiwanese-South Korean approach?

The United States is no stranger to top-down efforts, and many of them – the rapid mobilization during World War II, the race for the moon in the Sixties, even the Interstate highway system – have accomplished goals.

U.S. tech efforts this century, on the other hand, have been diffuse, and many people favor that multi-pronged approach. As Chris Fall of the Department of Energy's Office of Science put it to the *Washington Post*, "The beauty of how we do science in this country is that it isn't top-down."

When it comes to basic research, that model can work. When it comes to applying research, however, it might not. For instance, allowing industry to take the lead in the 5G area is producing disaster. There is, as noted, no American company competing with Huawei, which President Trump in August 2019 labeled "a national security threat."

An ideal solution would be for some American firm to buy either Ericsson or Nokia so that it would become an American national champion to confront China's, but the Federal government cannot force a private-sector business to do so. Moreover, it is of course not possible within the context of the U.S. system for the government itself to make such a purchase.

The Trump administration, however, has been thinking of alternatives, such as injecting cash into the Swedish and Finnish telecom-equipment giants and encouraging American companies Oracle and Cisco to enter the radio-transmission market. Moreover, the Defense Department wants U.S. businesses to develop open-source 5G software so telecom

---

At the moment, there is no American telecom-equipment giant, no "American Huawei."

---

carriers can buy off-the-shelf equipment and therefore bypass Huawei. American mobilization may also include subsidies to bring back high-tech manufacturing to the United States. Senators have proposed spending about $1.25 billion to encourage 5G research and development and to subsidize purchases of non-Huawei equipment by foreign telecom operators.

Dominance of future tech requires even greater effort, however. There must, as an initial matter, be a national push in education in STEM – science, technology, engineering, and math – similar to President Eisenhower's sponsorship of the National Defense Education Act of 1958, a direct result of the October 1957 Sputnik launch.

Education, of course, addresses only the contests of the distant future. Something is needed for the next decade. "If I were advising the president, I'd tell him, 'Go to Rice University where John F. Kennedy announced that we'd be on the moon by the end of the decade,'" said David Goldman. "Announce a Manhattan Project program to dominate technology in 5G telecommunications, in quantum computing, in a whole range of game-changing technologies, and restore American technological leadership."

Now, four specific actions are necessary. First, Federal funding for basic research, although not presently insignificant, should be increased so that it at least matches Chinese levels. This is the arms race of our era, and the United States can prevail. The American economy produced $21.7 trillion of gross domestic product in 2019. China's National Bureau of Statistics reported $14.4 trillion for that year, surely an exaggerated figure. Moreover, the U.S. economy is in reality growing faster than China's so there should be room in the Federal budget to out-research China.

Second, the Trump administration should, one way or another, get the FTC to drop its suit against Qualcomm in Federal court. In August 2019, the U.S. Court of Appeals for the Ninth Circuit, citing national security interests, questioned the Commission's suit against the company. Earlier, both the Defense and Energy Departments in court statements expressed concern over the FTC's action.

Nonetheless, the Commission is continuing to push its case. In the race for 5G, it's either Qualcomm or Huawei. Of course, the FTC should give no company a pass for clearly anticompetitive behavior, and competition law, like other law, should evolve, yet the Commission should not embrace such disruptive precedents at a moment like this.

At this moment, the national security of the United States, whether Americans like it or not, rests on the success of Qualcomm. Many may wonder how the U.S. ended up so dependent on one firm, but now is not the time to fix that problem.

Third, President Trump should continue the attack on Huawei. The administration has, to its great credit, sought the extradition from Canada of the company's chief financial officer – and Ren Zhengfei's daughter – for various crimes, but Meng Wanzhou, armed with the best lawyers China can buy, is unlikely to see the inside of a Federal prison anytime soon.

Huawei, however, remains vulnerable. There are bans on the purchase of its equipment, but more important are restrictions on sales and licenses to the company. The company and dozens of its subsidiaries and affiliates were added to the U.S. Commerce Department's Entity List beginning in May 2019. The designations mean no American business, without prior approval from the department's Bureau of Industry and Security, is allowed to sell or license to Huawei or listed organizations products and technology covered by the U.S. Export Administration Regulations.

Commerce, apparently acting at the direction of President Trump, had granted blanket 90-day waivers from the designations, meaning there were, as a practical matter, few restrictions on licenses or sales. Now, it has also been issuing approvals to specific companies for the sale of enumerated products, like memory chips. The waivers and approvals allowed Huawei to escape what was effectively a "death sentence."

Huawei, at least for now, cannot get by without American semiconductors and software. The company has said it has developed an operating system to replace Microsoft's Windows, and although in March 2019 Hua-

China has been called a "Subatomic Superpower."

wei said it would prefer to stay with the U.S.-based giant, it might be able to sell laptops without the world's dominant operating system.

Huawei, however, will have difficulty marketing phones without licenses from Google. The Shenzhen-based company unveiled its replacement HarmonyOS in August 2019 and later that month announced that its flagship Mate 30 phones would not come with popular Google apps such as Google Maps and YouTube and will not be able to access Google's Play Store. In short, the phones will not have Google Mobile Services.

Therefore, outside China, Huawei's phones, without a license from Google, will be extremely unattractive. Apart from Apple, no company – not even Microsoft, despite great effort – has developed a commercially viable ecosystem, because app developers have not provided support. Said analyst Richard Windsor to Reuters about the new Huawei phone line, "Without Google Services, no one will buy the device."

Because smartphones account for almost half of Huawei's business, the end of access to Play Store and Google apps could mean, as a practical matter, the end of Huawei as a profitable, stand-alone business.

On networking gear, the interruption of the supply of U.S. semiconductors would be serious. Huawei, in anticipation of friction with Washington, stocked up on at least three months' worth of chips according to Bloomberg. Nonetheless, the company would, in the event the Trump administration cut it off, run out of chips before it could develop its own or even find replacements.

China currently makes 16.3 percent of its requirement for chips, according to Brandon Weichert. Some estimates are higher, and eventually the percentage will increase, due in large part to the Made in China 2025 initiative and related efforts. As Claude Barfield of the American Enterprise Institute points out, Beijing is already "hell-bent" on developing its own semiconductor industry.

Weichert, in comments to me, said it would take China no more than two years to develop a "reliable indigenous semiconductor capacity." Huawei's Ren suggests a little longer, stating in August 2019 that in the next three to five years he will create an "invincible iron army" to counter U.S. sanctions.

In the meantime, Huawei could go out and buy Japanese and South Korean replacement chips, but they would not, in many cases, be as suitable as American ones. Moreover, the Trump administration could – and should – pull out all the stops and lean on Tokyo and Seoul to cut off supply to Huawei.

Finally, Washington can close a gaping hole in its own sanctions regime by interpreting its Entity List restrictions to cover chips made off of American soil by U.S. companies. Bloomberg in June 2019 reported that American firms, like Micron Technology and Intel Corporation, had resumed sales of foreign-made components to the embattled Huawei. The Commerce Department is now pushing rules to end this practice.

So the Trump administration can effectively force Huawei out of business by cutting off licenses and chips. In August 2019, Ren, the Huawei founder, issued a shock warning when he wrote in an internal memo that the company was at a "live or die moment." President Trump can now end Beijing's aspirations to dominate 5G, and he can make sure that 6G, whenever it comes, belongs to America too.

The Trump administration, however, is trying to strike a middle ground on the Entity List restrictions, apparently bowing to American companies and their desire to continue selling parts and issuing licenses. In July 2019, Commerce Secretary Wilbur Ross echoed earlier administration comments when promising his department would only issue exemptions "where there is no threat to U.S. national security."

That sounds reassuring, but it is not possible to divide Huawei into threatening and nonthreatening components. Huawei management can take profits from innocuous-looking parts of the business – Are there any? – to support the obviously dangerous portions. Money is fungible, so the only safe course is to prohibit all transactions with the company.

Ross also implied that licenses would be granted for items available from other countries, saying "we will try to make sure that we don't just transfer revenue from the U.S. to foreign firms." At first glance, sales of those items appear non-objectionable, but the better course would be to get all suppliers to stop all sales and licenses, to ring-fence Huawei by sanctioning any business that sells or licenses tech to the Chinese company.

Making companies choose between America or China would severely disrupt Huawei's operations, impede its current progress, and perhaps even force it out of business. In short, Ross could be underestimating America's leverage. The Commerce Department's approvals for U.S. chip makers to sell to Huawei, therefore, are short-sighted and should be reversed.

Fourth, there is an immediate need for restrictions on tech-sharing. The maintenance of research contacts with China is an extraordinarily complicated matter, especially in fields where America is ahead of China, such as quantum computing and various aspects of artificial intelligence.

Of course, there's no question that closing American facilities in China will inhibit, to some degree, American AI work. As Jeffrey

---

Xi Jinping does not accept that system of sovereign states competing in a network of treaties, conventions, rules, and norms.

Ding, the Oxford expert, points out, American AI outposts in China allow America to attract China's best researchers and to have "a listening post, an absorption channel of sorts, a way to be kept up to date on the Chinese ecosystem."

Despite the benefits of conducting research in China, the weight of evidence argues for closing American operations in that country. Unfortunately, the labs of American companies leak U.S. learning. Like water finding an equilibrium level, the net flow of technology is out of the U.S. into China.

Moreover, Chinese researchers, if they could not work for American companies in China, would not necessarily find employment in their homeland. Some of those seeking research slots would follow other Chinese to the United States, and some of those Chinese will, in all probability, become Americans. That outcome would exacerbate Beijing's brain drain. The U.S., by maintaining openness, can make that crucial problem even more severe.

The price of welcoming Chinese researchers to the U.S. is that some will come to America with the intention of going back to China or surreptitiously taking tech while in the U.S., but there is no way to prevent all of that leakage. Americans have to be confident enough that their system will be far more attractive to these researchers than a China returning to totalitarian rule.

Moreover, Americans cannot ignore the moral considerations of helping a militant, racist state use their work to support especially disturbing goals.

As questionable as, say, Google's decision to maintain substantial AI operations in China appears, the matter is ultimately not a Google issue. Corporate executives, in the final analysis, are not charged with defining and implementing America's national security policy.

This is a U.S. leadership issue. Until the American political system identifies China as an enemy or merely a foe – something that should have been done some time ago and which President Trump is on his way to doing – Google and other American businesses are free to work for Beijing, however wrong that cooperation could be.

The way to get American firms to stop working for China is to make it illegal to do so. Trump could declare an emergency pursuant to the International Emergency Economic Powers Act of 1977. Any such declaration could and almost certainly will be challenged in court, so legislation is ultimately needed. That's the job for 535 other individuals who are often found in Washington: 100 senators and 435 members of the House of Representatives.

AMERICA AS BRAZIL

It's the race of the century, and the U.S. urgently needs to improve the odds by going after China as hard as possible.

American companies look like they are firmly in China's court. They are, points out D.C.–based trade analyst Alan Tonelson, "voluntarily transferring defense-related cutting-edge knowhow to entities unmistakably controlled by Beijing, investing in Chinese tech start-ups and so-called venture capital funds, and building up their R&D and manufacturing capabilities in China despite intellectual property theft." America,

In May 2019, People's Daily *and Xinhua News Agency carried a piece declaring a "people's war" on America.*

he argues, "needs much more sweeping restrictions on American business operations in China."

Unless those sweeping restrictions are imposed soon, the export profile of the United States will resemble, as some say, Brazil's. Yes, America will be an exporter of primary products and a purchaser of manufactured and high-tech ones. The U.S. can avoid becoming a third-world economy, but leadership in Washington is absolutely essential.

Americans are not only in a fight they could lose, as experts say, but they also are, in crucial respects, already losing.

"Losing," however, does not mean "lost." The U.S., after all, was losing the race to the moon. The Soviets were the first to put a satellite into orbit, the first to put a human into space, and the first to conduct a spacewalk. Yet American drive, determination, and commitment meant that, 50 years ago, the first to walk the moon was an American.

So far, no other nation has left Earth's orbit.

First American edition published in 2020 by Encounter Books, an activity of Encounter for Culture and Education, Inc., a nonprofit, tax exempt corporation.
Encounter Books website address: www.encounterbooks.com

Manufactured in the United States and printed on acid-free paper. The paper used in this publication meets the minimum requirements of ANSI / NISO Z39.48–1992 (R 1997) (*Permanence of Paper*).

FIRST AMERICAN EDITION

LIBRARY OF CONGRESS CATALOGING-IN-PUBLICATION DATA IS AVAILABLE

SERIES DESIGN BY CARL W. SCARBROUGH